AF731068

Cells

Exploring the Building Blocks of Life

By Valarie Cevantes

Illustrated by Riki Artana

First Edition

Contents

Chapter 1: The Basics

Grab a pencil. Sharpen it to a point. Make the tiniest mark that you can on a piece of paper, so you can just barely see a little dot. Around 10,000 human cells can fit on that spot!

What are cells? Everything that is alive is made of cells. Humans are made of human cells. Plants are made of plant cells. Frogs, cows, butterflies, cats, and fish are all made up of cells.

The word **organism** means a living thing. Animals, humans, plants, and bacteria are all organisms. Some organisms, such as bacteria, are only made up of one cell. These are known as single-celled organisms. Other living things are made up of many cells, which are called multicellular organisms. All plants and animals, including humans, are multicellular because they are made up of hundreds of different types of cells. Your brain is made up of brain cells. Your blood is made up of blood cells. Even your bones are made up of bone cells!

Cells are often called "the building blocks of life". If you are building something using blocks or Legos, you need different shapes, sizes, and colors to make your creation. The more types of bricks you have, the more complicated the type of structure you can build. Cells are similar because they also come in all different shapes and sizes. When you put them together, they create an organism. The more types of cells you have, the more "complicated" the organism is. For example, a dog has more types of cells than a worm. If you were building a Lego car, you would need a block with wheels for it to be able to roll. Cells are also specially designed for the organism that they make. For example, cats have fur-making cells that humans do not have.

Fun Fact

Humans are made up of about two hundred different cell types. The human body has about thirty trillion cells in all!

Most cells are smaller than the thickness of a piece of hair, so they can't be seen with just your eyes. You must use a microscope to be able to see them. A microscope is a special tool that scientists use to get a closer look at the things they study. If you are taking a picture with a phone or camera, you can zoom in to enlarge an object. This is exactly what a microscope does, only it "zooms in" much more than your phone can. A microscope also has focus knobs so that everything comes in more clearly. If something looks fuzzy when you are trying to take a picture on your phone, you can click on the object that you want to come in clearer. This is what it means to focus on an object. Most phones can zoom in to about 10 times. The lowest magnification you need to be able to see common cells, such as bacteria and blood cells, is about 400 times. However, most scientists who are studying cells are using a 1,500 times magnification or more.

Cells must be small in order to do their jobs. For example, one of the jobs of a blood cell is to carry oxygen to different parts of the body. It is small and round so that it can quickly get to all of the areas of your body, from your brain to your fingers to your toes. If blood cells were big, it would be hard for them to travel fast enough to give every part of your body plenty of oxygen. If your organs don't get the oxygen they need, it could lead to you being tired and dizzy, having a hard time catching your breath, pain in your bones and muscles, cold hands and feet, and other issues that would make life uncomfortable. Luckily, all of our cells are designed for the jobs they do so that our bodies can function properly.

Fun Fact

Cell size is measured in microns. One micron is one-thousandth of a millimeter!

You might be thinking, how can living things be made of only cells? What about bones and organs? Well, these are also made of cells. **Tissues** are made up of similar cells that look alike and work together to do the same job. For example, your skin is a type of tissue. Your skin looks and feels like it's one big, smooth piece. However, if you used a microscope, you would see that your skin is just hundreds and hundreds of tiny flat skin cells that are stacked on top of each other!

Organs are just similar types of tissues working together. For example, your heart is an organ, made from three different types of tissues: muscle, connective, and epithelial. The muscle tissues in your heart work to pump blood in and out. The connective tissues connect the heart muscles to each other and help support the heart's shape. The epithelial tissue forms the lining of the heart and allows blood to move smoothly. These three tissues work together to form your heart.

Other common organs found in animals include the brain, lungs, liver, and stomach.

Lastly, you have **organ systems**, which are similar types of organs working together for a common purpose. For example, your respiratory system is the organ system for breathing. The organs that make up your respiratory system are your nose, mouth, throat, voice box, windpipe, airways, and lungs. The cells that make up these organs work together to allow you to breathe in and out. If one of these isn't working, like when you're sick and your nose is stuffy, it is more difficult for you to be able to breathe.

In summary, an organ system is made up of a group of organs working together. Each organ is made of groups of tissues working together. Each tissue is made of groups of cells working together. So when you break it down, every part of every living thing is really just cells working together!

Organisms are made of cells, but what are cells made of? Cells are made up of molecules, and molecules are made up of atoms. **Atoms** are the most basic elements of the universe. Anything that takes up space, alive or not alive, is made of atoms. You can find all of the different types of atoms on the periodic table, such as oxygen or iron. **Molecules** are atoms that are bonded together. For example, two hydrogen atoms and one oxygen atom bonded together create the common molecule H2O, otherwise known as water. Molecules and atoms by themselves are not alive. Cells are the smallest thing that can be alive, and cells are made up of many atoms and molecules! A cell is about ten times larger than a single atom.

A special molecule that most cells have is called DNA (Deoxyribonucleic acid). This is the master code that makes all living things what they are. The DNA in a plant cell decides things like how big its leaves will be. The DNA in a zebra cell might decide what its stripes will look like. DNA in a human cell might decide features like height or eye color.

Scientists use what's called **cell theory** to describe cells. This theory has three parts. The first is that *all living things are made up of one or more cells.* Something that is not made of cells, like your refrigerator, cannot be a living thing. Next is that *cells are the basic units of life.* They are the smallest living thing, and all life comes from cells. Lastly is that *all cells come from other cells.* This means that something that is alive cannot come from something that is not alive.

How are cells created? For around 2,000 years, scientists thought that life could be created through "spontaneous generation". This means they thought a living thing could come from a nonliving thing. For example, people noticed that maggots often appeared on rotten meat. It was widely believed that rotten meat "created" maggots. Today, we now know that life can't just be created out of nothing, and maggots actually come from eggs that are laid by flies. The reason we see maggots so often on rotting meat is because flies are drawn to rotten food, so they are more likely to lay eggs there. This goes with the third part of cell theory that all living things, including cells, come from other living things. Most cells come from other cells through a special process called *mitosis,* where the cell splits and divides in two.

Why do cells matter? Scientists study cells to learn how organisms function. The study of cells is called cell science, also known as cellular biology. They are figuring out ways to use this information to improve the quality of human life. For example, some scientists are discovering ways to use their knowledge of cells to create better food crops. This could help solve world hunger and help farmers to have bigger and healthier crops. Others are studying cells to cure diseases in animals. Medicine and other treatments are created by scientists studying the effect that they have on cells in the body that are making you sick. Learning about cells has also created new fields of science such as biotechnology, bioengineering, and biomedicine.

Chapter 2: Cells are Alive

You are alive. A cat chasing a mouse is alive. A fish swimming through the ocean is alive. It's easy enough to tell that animals are living things. What about plants? We know that they are alive, but they don't move around and breathe the same way that humans or your pet dog does. What about cells?

Scientists have created a system to decide if something is living or not. In order for something to be considered a living thing, it must be able to do ALL of the following: gas exchange, need water to survive, use energy, eliminate waste, respond to the environment, grow, and reproduce.

For animals, **gas exchange** is breathing. When humans breathe in, they are taking in oxygen and when they breathe out, they release carbon dioxide into the air. Plants do the opposite and take in carbon dioxide while releasing oxygen. This is why the air is cleaner when there are more plants around – they take what we don't need and turn it into what we do need! This is also why it's important that we plant trees and do our part for the environment. Plants help "clean" the air for us.

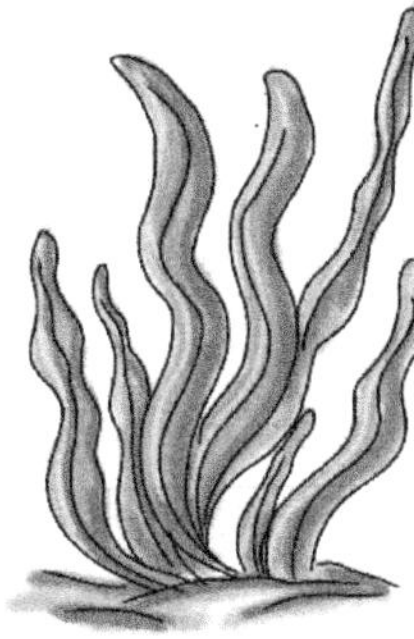

Fun Fact

If you are allergic to pollen and/or mold, there are some indoor plants that you can have that help reduce your allergies! NASA (National Aeronautics and Space Administration) conducted a study and found that there are quite a few specific plants that are good for people with allergies, including peace lilies, golden pothos, and dracaena plants.

For cells, gas exchange happens through diffusion across cell membranes. In other words, gas molecules flow in and out of the cells in a process called osmosis. These molecules are smaller than the cell membrane, so they can squeeze right through. This is like when you strain a pot of spaghetti. The water gets through because it is smaller than the holes in the strainer, but the spaghetti stays in because it is bigger. In the same way, molecules can get through the membrane of the cell.

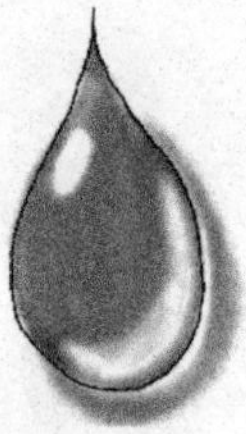

Plants, animals, and even cells **need water to survive**. Animals, including humans, often get water from the food they eat. Humans can only go about three to four days without water before they will die.

Fun Fact

The animal that can survive the longest without water is the Kangaroo Rat, who is able to go up to five months without water. It gets most of its water from the seeds it eats. It can actually go its entire life by getting water just from seeds!

The next requirement is that a living thing must **use energy.** For animals, this means we need to eat food. Our body uses the food we eat to generate energy. If you've ever missed a meal and felt extra tired, it's because your body needs food to give you energy. Plants don't eat food like we do, so they have to find another way to get energy. Their energy comes from the sun in a process called photosynthesis. Cells live by absorbing nutrients from the food that you consume. If you eat only junk food, your cells will only have junk food to live off of, resulting in both of you being unhealthy! Unhealthy cells lead to them not being able to do their job well, which can lead to illnesses.

Fun Fact

It takes cells longer to dispose of beta-carotene that is found in orange foods, such as carrots. If you eat a lot of carrots (about ten carrots every day for around a month), your skin can actually start to turn orange!

For animals, **eliminating waste** is when our body takes what we don't need from the food we eat and gets rid of it by using the bathroom. Plants get rid of their waste in the soil. Cells must also get rid of waste. They take what they don't need, break it down into a little "package", and dump it out. Another cell might pick up that waste and use it if it needs it, so the waste is recycled among the cells.

Fun Fact

Farmers often rotate what plants they grow each year. They may grow corn for one year, and then switch to beans the next year, and then corn the next. This is because the soil is always changing from the plant waste, and beans will grow better in the waste left by corn, while the corn will grow better in the waste left by beans!

Respond to the environment means there is a reaction based on the setting. For animals, this could be that we shiver when we are cold, sweat when we are hot, or shut our eyes when it's too bright. Birds going south for the winter is responding to the environment – when it is cold, they are going to where it is warmer. Some animals that live in colder climates grow more hair than animals that live in warmer climates. Plants need more water in the summer when they get more sun. Some plants have leaves that will change color or fall off in the fall when the weather gets colder.

Cells also respond to their environments. Cells can sense changes in temperature, chemicals, and other parts of their environment. Cells on the skin respond to touch and temperature. Cells in the ear react to sound. Cells signal the brain when it senses changes. For example, have you ever felt yourself getting sunburnt? If your skin cells sense that this is happening, they will signal the nerve cells to send a message to your brain to warn you that your skin is getting too hot.

All living things must **grow** in some way. Humans and animals grow from a baby to an adult. Some animals grow a very small amount, some a very large amount. Plants grow from a seed to a plant. Cells also grow; some grow a lot and some only grow a little. Hair follicles, skin, and the blood cells are some of the fastest growing cells in the human body. It makes sense if you think about it – you have hair that falls out every day, but you don't go bald. You can get sunburnt or have a cut in your skin, but your body regenerates new skin cells quickly to replace the damaged ones that flake off from the sunburn or that you pick off in the form of a scab. There are people who donate blood regularly, but their body makes more to replace it.

Fun Fact

Kangaroos are one of the animals that grow the most from birth to adulthood. They are technically only about the size of a bean when they are born. Then they live in their mother's pouch for around a year, and pop up as a little baby kangaroo, otherwise known as a joey!

All living things must **reproduce**, otherwise they will go extinct. For humans and animals, reproduction comes from having a baby. For plants, reproduction comes from seeds. Cells have a special kind of reproduction called mitosis, where they grow double of everything and then split in half to form two cells out of one.

Animals, plants, and cells are alive because they do ALL seven of these things. This means that you as a living human have at least seven things in common with any other living thing. You have these seven things in common with a sunflower, an elephant, a tiny speck of bacteria, and even a mushroom. Who would have thought that you have seven things in common with a mushroom??

Something that was once alive but is not alive anymore is called **dead**. A fallen tree is dead. Albert Einstein is dead. The blueberries in your fridge are dead. Something that was never alive is called **nonliving**. An airplane is nonliving. Clothes and shoes are nonliving. They were never alive, so they can't be dead. What about a chair that is made from wood? Wood is alive, so is the chair dead, because it is made from dead wood? No. A chair is nonliving because the chair itself was never alive. The chair was never walking around town. You also don't see chairs growing in anyone's garden. Wood that has been manipulated or changed to be turned into a chair is different than a tree that was once alive and is now dead.

Is fire alive? At one point, it was believed that fire is a living thing. It **exchanges gas** because it needs oxygen and it releases carbon dioxide. It **uses energy** because it needs wood to survive. It **eliminates waste** by releasing smoke and ash. It **responds to the environment** by getting smaller when it is cold/raining, and it moves with the wind. It **grows** by getting bigger when it has more wood. It can **reproduce** if you have a match and light the match with another match, or light a piece of wood with one that is already on fire. However, it doesn't NEED to reproduce to not go extinct – fire does not only come from other fires. Also, it does not need water to survive. Fire only has five of the seven characteristics of a living thing, so it is nonliving.

Chapter 3: DNA, Genes, and the Life-Cycle of a Cell

You may have heard of DNA in the courtroom, something like, "the DNA evidence proves that this man did the crime!" Police can catch criminals using DNA tests. They can get a person's DNA from a bit of hair, saliva (spit), or blood.

What is DNA? DNA are special molecules that contain the instructions for life. It is like a code that tells cells what to do. DNA is what makes you a person and not a fish, squirrel, or tree. It also determines what you should look like and even parts of your personality. It is a very long chain of information formed with short pieces called genes.

The nucleus of a human cell has around 25,000 genes. They carry the traits we inherit from our parents. One gene is just a section of DNA that tells the cell to make a particular protein. These special proteins determine features such as hair texture and foot size, and if we even have hair or feet at all! They also make sure that skin grows and blood flows. One gene might produce a single trait, like the ability to roll your tongue. But genes usually work together.

Animals and plants also have genes. About 12,000 years ago, humans started taming wild animals. They discovered that if they breed fast horses with other fast horses, they could "create" even faster horses. The same thing worked with plants. Apples, corn, potatoes, rice – every plant that gives us food today has been combined with others to be bigger or more useful. Humans have been doing this for years but no one could explain how this process worked. In the late 1800s, scientists discovered heredity, or the passing of genes from generation to generation. Genes are the most basic unit of heredity.

You may have had someone tell you that you "get your eyes from your mom" or "you have your dad's sense of humor". These traits that you have in common with your parents are an example of the genes you inherited from them: heredity. This was an important discovery for scientists because they found that other things, such as diseases, can be passed down through genes as well. When you go to the doctor's office, they will often ask for your medical history to see if your parents have a disease or condition that they should be checking you for. For example, there are some cancers, poor eyesight, and some types of diabetes that you are more likely to have if your parents also have it.

Another part of your DNA are chromosomes. You may have seen pictures of chromosomes, which look like a twisted ladder. Each human cell contains 23 pairs of chromosomes, or 46 chromosomes in all. Cells have 46 chromosomes, which are made of DNA, which are made of genes.

Normally when you think of division, you think of numbers getting smaller. With cells, division is how they grow! We know cells must reproduce so that they don't go extinct. They do this by dividing – a single cell will split into two separate cells. There are two different ways that this can happen; mitosis and meiosis.

Most cells in the human body go through mitosis. This is when one cell divides into two identical cells, like a clone. When a cell is preparing for mitosis, it will "grow" double of everything. Once it has double of everything, it splits off into another cell, and now the one big cell has become two cells that are exactly the same. How crazy would that be for a person? You would suddenly grow another set of arms, legs, you'd have two heads, and so on until finally you have two of everything and you just split perfectly in half, so now there are two identical versions of yourself!

The only cells that go through *meiosis* are reproductive cells to create a baby. They duplicate and then split twice, creating four different half-cells. They all have 23 chromosomes instead of the normal 46. These special 23 chromosome cells are called zygotes. For a baby to be created, a female zygote combines with a male zygote. This means half of the chromosomes and DNA come from the mom, while the other half come from the dad. This is where hereditary comes in – each person is unique because a baby gets half of their DNA chromosomes from their mom, and the other half from their dad.

> ***Fun Fact***
>
> *The eggs of most animals are actually one giant cell. One of the largest cells is the ostrich egg! It can weigh over three pounds.*

Sometimes a baby has an extra copy of one of those chromosomes and ends up with 47 instead. This is a condition called Down syndrome. No one knows for sure why this happens. Down syndrome occurs in about 1 in every 700 babies. It is a lifelong condition that may include lower intelligence, speaking slower, and many physical features such as poor muscle tone, being shorter, a flattened face, and small ears, hands, and feet. It might take a baby with Down syndrome longer to talk and walk. Down syndrome varies from person to person, and most people who have it are able to attend school and have many great achievements.

Fun Fact

Melissa Reilly is an athlete with Down Syndrome who has won the gold medal in skiing, swimming, and bicycling at the Special Olympics! There are also many professional actors with Down Syndrome, including Chris Burke from Life Goes On; Jamie Brewer from American Horror Story; and Lauren Potter from Glee.

A baby starts as a single cell, formed from two zygotes (one from its mom and one from its dad). That cell goes through mitosis to divide and divide until it forms a tiny ball of cells. This is called an embryo. It continues to do this until it gets bigger and is called a fetus. Eventually it continues to divide into specific cells with specific jobs, such as kidney cells to create a kidney, blood cells and more, until it has all of the parts to be a functioning baby!

Fun Fact

Babies are born with around 300 bones. Adults have about 206 bones. What happens to all of those bones? Babies have more because their bones still need to grow and will eventually attach together.

Most types of human cells reproduce only themselves. For instance, muscle cells can only make more muscle cells. However, there is a special type of cell, called stem cells, that can change into other types of cells. This allows them to act as a repair system for the body. If you have kidney disease and your body needs more healthy kidney tissue, bone marrow stem cells can reproduce into the kidney tissue your body needs!

All cells will eventually die. Some of our cells are constantly dying, and then being replaced. For example, we shed millions of skin cells every day. Your body is continually growing and replacing your skin cells. This is especially true if you have dry skin or if you are sunburnt. Your skin cells flake-off of your body and then get swapped for new ones.

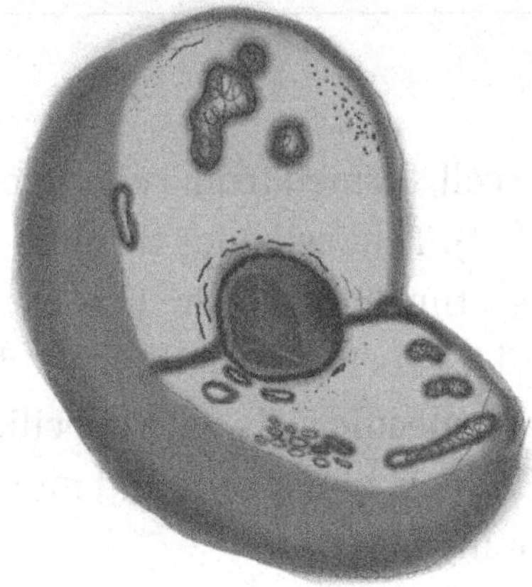

Here is a life span of some cells found in the human body:
Skin cells: 1 – 34 days
Stomach lining cells: 2 days
Red blood cells: 120 days
Bone cells: 25-30 years
Nerve cells: Live forever until we die

Fun Fact

Nerve cells are the only cells that we never replace! The nerve cells we are born with grow, but don't die until we do.

People who donate blood give their healthy blood cells to people who need it. Someone may need it if they have been in an accident where they lost a lot of blood and their body cannot replace it fast enough. Other people may need healthy blood if they have a disease in which their blood cells are not healthy or are dying quicker than their body can make more. If you do donate blood, the nurses or doctors are only taking about half a liter so that your body is able to make more to replace it. This is why you usually can't donate blood more than once every two months.

Fun Fact

The average person has about 1 ½ gallons of blood in their body!

You can't donate blood to just anyone though; you have to have a compatible blood type. There are four main blood types: A, B, O, and AB. These letters stand for the type of protein that is found on a person's red blood cell. If someone only has A proteins on their cell, they have type A blood. People with only B proteins on their red blood cell have type B blood. People with both A and B proteins have type AB blood. This is the rarest blood type, and only about 4% of people in the United States have it. Some people don't have any proteins on their red blood cells. They have type O blood. Type O blood is considered the universal donor because anyone can receive type O blood. Type A blood can only be received by people who have type A or type AB blood. Type B blood can only be received by people who have type B or type AB blood. Type AB blood can only be donated to other type AB blood (but they can receive it from any other blood type). Type O can only receive from other type O blood (even though they can donate to every other blood type).

Chapter 4: The Guts of the Cell

How would you describe a house? Rooms have walls, windows, doors, and furniture. There are many parts to a house. Cells are also made up of many parts. Instead of walls, cells have cell membranes (which means "thin skin"). In your house, you might have an office for you to do your homework. A cell has a nucleus where all of their "paperwork" is done and important information is stored. In your house, you might have electricity that provides all the power – giving it light and heat. A cell has mitochondria, which powers the cell. A house has all kinds of different rooms with different jobs, just like a cell has different organelles with different jobs that make up a cell.

All animal cells pretty much look the same, whether they come from a human, elephant, fish, or bird! Animal cells have the following organelles:

- Nucleus (new – klee – us)
- Cytoplasm (sight – oh – plaz – um)
- Ribosome (rye – boe – s"oh"me)
- Golgi Apparatus (goal-dgee app-uh-rah – tus)
- Mitochondria (my – toe – con – dree – uh)
- Two types of Endoplasmic Reticulum, usually referred to as ER
  - Smooth ER
  - Rough ER
- Vacuole (vah – cue – ole)
- Cell Membrane
- Lysosome (lie – so – s"oh"me)

- Plant Cells have all those organelles except for lysosomes. They also have other organelles that animals don't have:
- Cell Wall
- Chloroplasts (klore – oh – plasts)

The **nucleus** is like the main control center of the cell. Most cells only have one nucleus. It contains and guards the DNA. It also makes sure that the cell is functioning as it should be. It is like the parents or adults of the house. In most homes, they are the ones that control what happens in the house and makes sure the house is working properly. It could also be compared to the office of a house, as it is the place where "paperwork" is done and all of the important information is stored.

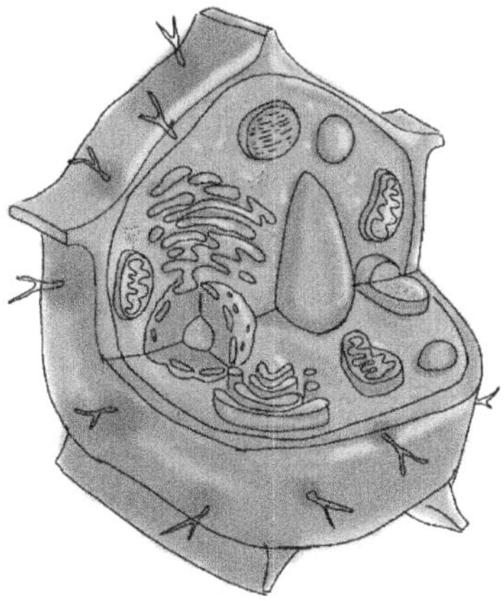

Fun Fact

The nucleus is usually quite large, and so it was the first organelle to be discovered when scientists were studying cells!

The **cell membrane** is the outside wall of the cell that lets in nutrients, lets out waste, and protects the rest of the organelles. It is like the walls and doors of a house. The walls of a house protect everything that is kept inside, while the doors are a gateway to let things in and out.

The **cytoplasm** is a jelly-like substance where the organelles float around. This is like the "blood" of the cell or the air inside of a house. It fills up the empty space around everything else. It allows the organelles within the cell to move around to do their jobs. Another way to think of this is if you have toys in a bathtub but no water, the toys stay in one spot. If you add water, the toys can move around or float. Cytoplasm is like the water in the bathtub, allowing the organelles to easily move around within the cell.

The **ribosomes** of the cell make and store proteins for the cells to use as energy. This is like the kitchen in a house. The kitchen is where meals are cooked and prepared. The kitchen also has a fridge and pantry to store energy or food for you when you are ready to eat. Ribosomes are one of the smallest organelles in a cell. Unlike a kitchen, however, a cell contains many ribosomes. There are some that are scattered around the cell to easily transport the proteins that it has made within the cell, and some that cling to the rough ER.

The **rough ER** is called rough because it's covered in ribosomes, which looks like little bumps. It is in charge of folding, sorting, and transporting proteins outside the cell to their proper destinations, especially the Golgi apparatus. This is kind of like cooks in the kitchen! They use what's in the kitchen (ribosomes) to prepare the "energy" or food and get it to the right place outside of the kitchen – like a plate of food to be given to you in the dining room.

The **smooth ER** isn't covered in ribosomes, so it looks like a smooth surface. Its job is to make the lipids (fats) that are used to make the cell membrane. This is like the garage or workroom of a house, because it might have materials to repair and strengthen the outside of the house.

Fun Fact

Your liver cells have more smooth ER organelles than the cells in other parts of your body! This is because the smooth ER also helps with removing toxins, which is a major job of your liver.

The **Golgi apparatus** is sometimes called the **Golgi body**. It takes the proteins from the ribosomes and assembles them into larger proteins. It also delivers proteins in and out of the cell. This is kind of like a mailbox. A mailbox is used to store and deliver letters/packages until they are ready to go to another part of the home.

The **mitochondria** is the powerhouse of the cell! It breaks down food into energy for the cell to use. This is like the electricity in a house. Electricity supplies a house with energy to be able to function, such as lights, heating, and air conditioning. The mitochondria supplies energy for the rest of the cell.

Fun Fact

Different types of cells have different amounts of mitochondria. Cells that need a lot of energy, like your muscle cells, will contain more mitochondria!

The **vacuole** is used to store the materials needed by the cell. They are smaller in animal cells, and larger in plant cells. It is also used to store waste until there is time to remove it. This is like the fridge or pantry of a house. They store food that you need until you eat it, or the food goes bad and needs to be removed.

The **lysosome** is only found in animal cells. It works as the digestive system of the cell to break down enzymes and discard cell waste. Plants don't need lysosomes because they don't eat food like animals do; they get their energy through the sun in a process called photosynthesis. The lysosome is like the garbage disposal in a house. It grinds up and gets rid of unwanted materials.

The **cell wall** is only found in plant cells. It goes outside of the cell membrane as an added layer of protection against things like insects. It also gives plants the structure they need to stand tall. This is like the walls of your house because they give a house support and protect it from things outside.

The **chloroplasts** are found only in plant cells. This is what the plant uses instead of lysosomes. Chloroplasts absorb energy from the sun and uses it to make energy or food for the plant in a process known as photosynthesis. This is like the kitchen of the plant cell, as it uses materials to create energy or food for the plant.

Fun Fact

Chloroplasts also contain chlorophyll, which is what makes plant leaves and stems green!

What are organelles made of? Just as your house is made up of wood, metal, plaster and tiles, organelles are made of a variety of materials. These include water, proteins, carbohydrates, nucleic acids, and fats. Water makes up about 90 percent of a cell's weight. Protein molecules make up about 5 percent. Carbohydrates are the sugars, which are burned for quick energy. They make up about 2.5 percent of a cell's weight. Nucleic acids include DNA, which is found in the nucleus. They make up about 1.5 percent of cells. The last 1 percent is found in the cell's outer membrane, which is made up of the lipids (fat and oil) that are made in the smooth ER.

You may have noticed that a lot of these organelles work to create and distribute proteins! There is a good reason for that – about half of the stuff in your body is made of protein. Each person has about 500,000 different proteins working at any time. Hair follicles (the tiny holes in your skin and head where your hair grows out of) and nails are also made up of protein. Your hair is made up of a special type of protein called keratin.

Proteins have a few different functions in our body. Some proteins help cells keep their shape or make up muscle tissue. This is one of the reasons that people who are very serious about working out will drink protein shakes. They need extra protein for all the new muscle tissue they are creating! A lot of proteins work as switches that turn chemical reactions on and off when needed. For example, some proteins are responsible for metabolism. Metabolism is when our food is changed into energy. Some proteins work as antibodies. These proteins help us when we are sick by attacking the germs or bacteria that are making you sick. Another common type of protein is called an enzyme. Its job is to help the cell get things done quickly. Enzymes are created for very specific purposes – if the mitochondria need help breaking down food into energy, your body might create special enzymes to help the mitochondria speed up its job. These enzymes can be used over and over!

When you have a minor ache or pain, like a headache, you might take aspirin. The ingredients in the aspirin work to block one of your enzymes that works to create inflammation, swelling, pain, and fever. It can stop your body from feeling pain and make your fever go away, because it's stopping those proteins in your cell. It can also stop blood clots by breaking up the blood that is clotting, which is great for your arteries or if you are about to have a heart attack. You have to be careful not to take aspirin unless you really need it though, because sometimes you need your blood to clot to stop you from having too much blood loss. Your stomach can be especially sensitive to too much aspirin and the cells in the lining of your stomach could be permanently damaged if you take too much or too often.

Fun Fact

Some medicine has a bad taste to keep kids from taking too much of it, which can be dangerous for their health!

Chapter 5: Food for Your Cells

What happens to a piece of food when you eat it? Before you even put the piece of food in your mouth, the cells in your mouth start to prepare by creating saliva or spit. Your saliva helps to break down the food while it's still in your mouth. When you swallow, the food goes down your throat to your stomach in a tube called your esophagus (ih-sof-eh-guss). Once your food gets to your stomach, the cells that make up your stomach have three things to do. The first section of cells work as storage for the food that you've eaten. Another section of stomach cells work kind of like when you put food into a blender - they take what is in storage and begin to break down the food to be a mushy, liquid mixture. The stomach has acidic liquid in it already that mixes with your food to help break it down and help kill bacteria that might be in the food. Then the last section of stomach cells will slowly empty that mushy mixture into another organ called the small intestine.

Your small intestine is a long tube that is all scrunched up in your stomach. The job of the cells in this organ is to keep breaking down the food from your stomach and pick out all the "good stuff" like vitamins, minerals, proteins, carbohydrates, and fats for the rest of your body to use. Other organs such as the pancreas, liver, and gallbladder help with this extraction as well. These nutrients are sent to the blood, and your blood cells help to send them to the liver. The cells in your liver will filter out any waste and figure out how much of your nutrients will go to the rest of the body, and how much will stay behind as storage for your body to use as energy later.

Once the nutrients have been pulled out of the food you eat, anything that is left is waste. It is what your body can't use and will enter the next organ called the large intestine. It is similar to the small intestine in that it is also a long tube that is scrunched up in your stomach. The difference is that it is a wider and actually shorter tube! This waste passes through the large intestine as a last chance for your cells to get any water or minerals before it's ready to leave the body. As your cells extract water from this waste, whatever is left eventually turns solid. This waste goes by many names, such as poop, stool, or a bowel movement. The poop gets pushed to the very last stop called the rectum. This is like a storage area until you are ready to go to the bathroom and get rid of it by pushing it out.

You may have heard the saying, "you are what you eat". This is because your cells get their food from the food that you eat. The health of your cells, tissues, organs, and mind all depend on getting the stuff it needs from your food. If you only eat junk, eat the same food every day, or don't eat enough food, your body cannot get the energy it needs and won't work properly. So if you eat junk, your body becomes junky. If you eat healthy, you'll be healthy...you are what you eat. If you only eat grilled cheese and chicken nuggets for every meal, your body will get protein from the chicken and calcium from the cheese. However, it will be missing lots of other vitamins it needs. All the different types of cells in your body need vitamins, minerals, proteins, carbohydrates, and fats in order to function properly. It is important to be eating a bunch of different types of foods so that your body can get a variety of these important nutrients!

Some nutrients are found more abundantly in certain types of foods. Foods that are naturally red, like watermelon, tomatoes, and peppers;

contain an antioxidant called lycopene. This can help protect your heart and skin by helping protect from cancer, heart disease, and more! Foods that are naturally orange, like sweet potato, carrots, pumpkin, and mangos; contain a lot of Vitamin A. This vitamin is especially good for the cells in your eyes, heart, lungs, and kidneys to function properly. Foods that are naturally yellow (and sometimes a bit orange), like lemons, pineapple, yellow peppers, cantaloupe, and oranges; contain a lot of vitamin C. Vitamin C helps your skin cells to heal cuts and keeps the cells in your gums strong, to keep your teeth from falling out! Foods that are naturally green, like broccoli, spinach, green peppers, and peas; have prebiotics. Your stomach cells use prebiotics to kill off any bad germs that enter your body to help keep you from getting sick. Foods that are naturally blue or purple, such as radishes, purple cauliflower, blueberries, and grapes; have antioxidants in them. These are especially important for your brain cells, who use antioxidants as protection from damage over time to help with functions like memory.

Fun Fact

Have you ever heard that eating carrots helps you see in the dark? This is because carrots have a lot of something called compound beta carotene. This is a type of vitamin A that your cells will take from the carrots after you eat them and use to help the cells in your eyes adjust in the dark! So it's true, eating carrots DO help you see better in the dark!

While the cells in your body are absorbing all of your food and deciding where to put all the nutrients, they are also getting energy from your food. If you skip a meal, you may notice that you are extra tired or have trouble thinking/staying focused later. After you eat lunch, you may get a burst of energy or motivation. This is because your cells need food in order to keep going and give you energy. Our cells changing food into energy is a chemical reaction called **metabolism**. You use this energy to do everything, including physical activities like walking, mental activities like thinking, and even growing!

The process of metabolism occurs when your food is going through your small intestine and being picked apart for all of its usable vitamins and nutrients. Your cells will also send the sugars, amino acids, and fatty acids that are in your food as an energy source to other cells – especially the liver, muscles, and body fat. A simple way that many think of what metabolism is, is that it makes a difference with how quickly your body gains or loses weight.

You have probably heard of the word calories. Many people who are trying to lose weight will keep track of how many calories they consume in a day, and try not to eat too many. Calories are not actually a bad thing – they just measure how much energy that food will provide the body. In fact, some food labels (especially in places like Europe) will label their food with the word "Energy" instead of "Calories". A bag of chips has more calories than an apple, so the chips will give your body more energy than the apple. But having more energy, or calories, than what your body actually needs isn't always a good thing. If you eat or drink more calories than what you need, your body will store the extra calories for later. These calories are stored as fat and too much can cause your cells to have to do extra work to keep it all, which can eventually lead to a stroke, diabetes, or other health issues.

Your body burns calories all day long. When you walk, when you breathe, when you write, when your heart is beating, and even when you sleep. Everything you do requires energy, which means your body is constantly using these calories. When you exercise, the cells in your body need extra energy. They use the calories stored in your body fat to give your body the extra energy it needs. This is how exercising can help a person lose weight. If you are active and exercise regularly, as well as not eat too many extra calories, your body will have a faster/higher metabolism. This means that it will burn away calories even faster, even while you are resting.

This doesn't mean you should never eat junk food or foods that are high in sugar or calories. It just means that you should eat those foods in moderation, while also giving your cells a variety of other foods that contain the nutrients they need to grow and do their jobs. It's also important to remember that you don't want to limit how much you eat or how many calories you give your cells. If you don't give your body enough calories, your cells will not be able to reproduce and keep doing their jobs correctly. This can lead to you getting extra tired and unable to focus, and your cells will have a harder time killing bad germs so you are more likely to get sick. It can also lead to permanent problems like you not growing as tall as you should or would have otherwise. You might be extra hungry and need more calories than normal when you are going through a growth spurt, because your cells need that extra energy to do all of that extra growing.

Chapter 6: Animal Cells

What is blood? It's a liquid. It's red. If you get a cut or scrape, it can come out of your body. Blood is made up of two types of cells – red blood cells and white blood cells. They are both made inside of your bones! Red blood cells have the job of carrying oxygen around the body and helps to remove carbon dioxide from the body. A red blood cell looks like a small disk, and they live for about four months before being replaced by new red blood cells. The reason that your blood and red blood cells are red is because they contain a lot of iron.

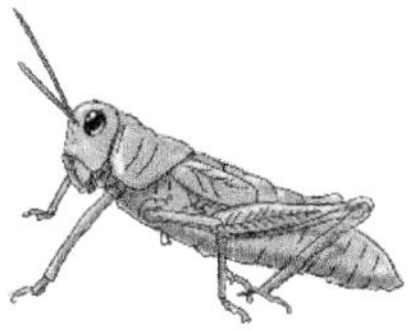

Fun Fact

Not all organisms have red blood. Grasshoppers and some lizards have green blood while lobsters, crabs, and snails have blue blood!

White blood cells work to keep you healthy. They travel through your bloodstream to look for bacteria, infections, or anything else that can make you sick. When they find something like this, they quickly attack it so that you can stay healthy and safe. If you are sick, your body will create more white blood cells to help fight off the infection. You have different types of white blood cells to help fight off different types of infections. For example, some fight off bacteria while others might fight off allergies. They are shaped like a bumpy snowball.

> ***Fun Fact***
>
> *There are around 7,000 – 25,000 white blood cells and about 5 million red blood cells in a single drop of blood!*

Your body has four different types of tissues: connective, muscle, epithelial, and nerve. *Connective* tissue is what your body has the most of. Its main job is to join together other tissues. It also surrounds and cushions organs, bones, and muscles and helps to hold them together. Examples of connective tissue include tendons, ligament, and cartilage. Your bones are also a special type of connective tissue. *Muscle* tissue is responsible for movement. This includes movement you control, like bending your elbows, and involuntary movement, like your heart beating. You have over 600 muscles in your body! Skeletal muscles are what you use to move your arms and legs. Cardiac muscles make your heart beat. Smooth muscles are all over your body – even in your eyes to help them focus! *Epithelial* (eh-puh-thee-lee-uhl) tissue makes up the coverings and linings of surfaces on the body as protection. For example, the lining of your stomach and the outer layer of your skin is made up of epithelial tissue. The last type of tissue is the *nerve*. This is made up of long, stringy looking cells called fibers that go throughout your whole body. Its job is to receive and send information throughout the body. So if you touch a hot stove, the nerve cells in your fingers send a message to your brain that says something along the lines of: "This is hot! It hurts! Stop touching the stove or else there will be permanent damage to the body!"

Fun Fact

Shivering is actually just hundreds of muscles contracting and relaxing to make heat to try to warm you up!

One of the most important organs in your body is your brain. The cells in your brain work together to control just about everything you do. This includes things you choose to do, like wiggling your toes or scratching your face. Your cerebrum is the biggest part of your brain. The cells that make up this part of your brain are responsible for controlling your muscles and helping you to think. It also controls how you talk, learn, your memory, and emotions.

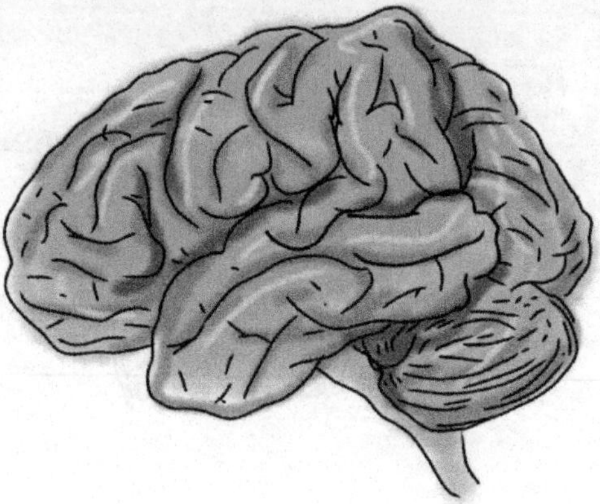

Fun Fact

The left half of your cerebrum controls the right side of your body, while the right half controls the left side of your body!

Another group of cells in your brain work together to form your cerebellum (s-air-uh-bell-uh-m). When you have different muscles working together, your cerebellum helps to align them. For example, when a baseball or softball player is swinging a bat, they don't just use their arms. They use their arms, abdomen, and legs to get a full body swing. As well as helping you balance, your cerebellum helps these muscles work together in one smooth motion!

Fun Fact

The word cerebellum is Latin for "little brain"!

The last part of your brain is a group of cells working together to make up your brain stem. It is located in front of your cerebellum and under the cerebrum, and is attached to your spinal cord, which runs all the way down the inside of your spine. The job of these cells is to control things that happen involuntarily in your body to keep you alive. This includes your heart beating, breathing, blood pressure, and digestion.

The human body is made up of ten different organ systems: the respiratory system, digestive and excretory system, circulatory system, urinary system, integumentary system, musculoskeletal system, endocrine system, lymphatic system, nervous system, and reproductive system.

Your musculoskeletal system is made up of your bones and muscles. The job of this system is to support your body, protect important organs from damage, and make blood cells. Your muscles and bones are connected to each other with tendons. They work with the nervous system to get messages from your brain to move, like lifting your arm or kicking your leg. Your bones are the part of this system that helps to protect your organs. For example, the bone cells that make your ribs protect your heart and lungs, and the bone cells that make your skull protects your brain.

Fun Fact
The smallest bones in your body are called ossicles, and they are located in your ear!

Your digestive system includes your stomach and intestines. The job of this organ system is to break down the food you eat, take nutrients from the food, and deliver them to the rest of your body.

Your cardiovascular system includes your heart and veins (otherwise known as arteries). The job of this system is to make sure that blood is pumped throughout your body. The cells in your blood carry oxygen and nutrients to other organs that need it.

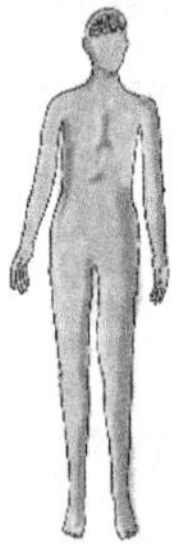

Your nervous system isn't about being scared, but all about sending messages! The organs that make up this organ system include your brain, spinal cord, and all of your nerve tissues. These work together to send messages throughout your body to keep it safe from the outside world. For example, if it is cold outside, the nerves in your skin will alert your brain so that you know that you should put a coat on!

Your circulatory system is responsible for moving things like oxygen, nutrients from your food, and waste throughout your body. When your cells are busy working to do their jobs to keep you alive and thriving, they create a waste called carbon dioxide. They have to get rid of it by sending it to your lungs, so you can breathe it out. When you breathe in, your lungs are absorbing oxygen from the air, which your blood cells will take to the rest of your body. The reason you breathe harder and faster when you exercise is because your heart is beating faster, so it needs more oxygen to keep beating.

Hair is actually on almost every part of your body. The job of hair is to help you from getting too hot or cold and to protect your skin from the sun. The cells in your mouth and nose work together to help you breathe and taste. Your nose can warn you from danger like smelling smoke from a fire. Smelling also helps with taste. This is why sometimes when you are sick with a stuffy nose, you may not be able to taste things as well as you normally can.

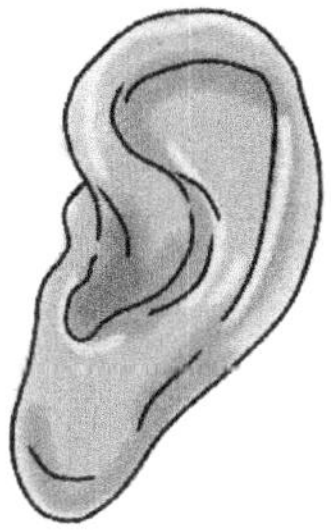

Fun Fact

In addition to hearing, your ears also help you to keep your balance!

One part of your mouth includes your tongue, which is where your taste buds are located. If you look closely at your tongue in the mirror, you may notice that it is covered in bumps. These are called papillae (puh-pill-ee) and is where your taste buds live. There are four basic tastes: *sweet, sour, bitter*, and *salty*. Some people think that different parts of your tongue sense different tastes, like the tip of your tongue has the most taste buds for sour. However, this is not true and your taste buds are actually sensed all over your tongue. Scientists have recently revealed another taste called *umami* (oo-mah-mee). It is mostly found in Japanese food, and has a savory or rich, meaty flavor. It can also be found in the taste of bacon. Some scientists also list the taste of *fat* as another flavor that can be sensed by taste buds. It is similar to a bitter taste, but falls under its own category. Something that is high in *fat* taste could mean that the food has gone bad, and is a warning to the body not to eat it. One of the reasons these two tastes can be separated into their own categories is because our body digests them differently. Your taste buds work as a signal to your brain for you to enjoy your food, but also as a signal to your body for how to prepare to digest it.

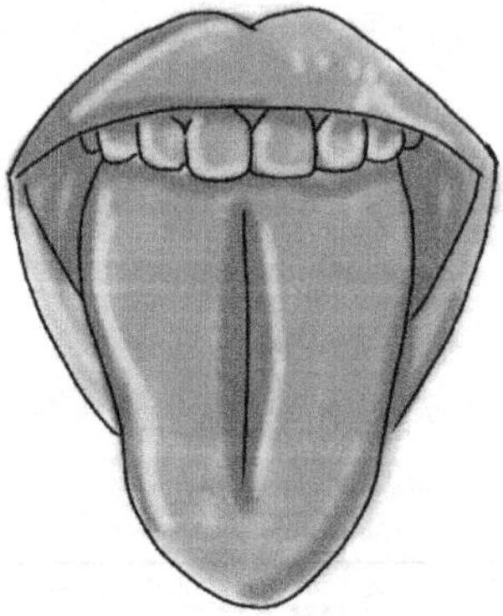

Fun Fact

Most people have about 10,000 taste buds at one time! If you burn your tongue on hot food or drink, you can kill off some of your taste buds. It won't last long though! They have a life-span of about two weeks, so they are constantly replacing each other.

Chapter 7: Sickness and Cell Irregularities

Your body works hard to keep you safe and healthy. The first barrier on your body to keep out germs is your skin! The job of your skin cells is to work as a barrier to keep out anything unwanted from the rest of your body. Your nose hair is also a line of defense. It works like an air filter by trapping germs that try to get in through your nose. Mucus, or boogers, do this as well. If germs do make it inside of your body, your white blood cells go to work to destroy them. This can take many days, which is why it sometimes takes a few days of feeling sick before you get better. Your body makes more white blood cells to help fight off these germs so that you feel better. Sometimes the germs or sickness can be too much for your white blood cells. If this is the case, there are special proteins in the blood called antibodies that will attach themselves to the germs as a way to learn how to stop it. It will then signal other antibodies to come in and help fight off the germs.

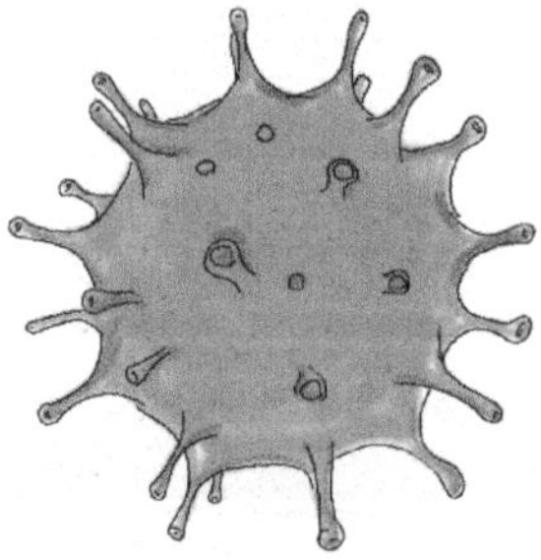

Fun Fact

Your body has something called immune memory. This means that if the exact same type of germ or sickness enters your body, it already knows and remembers how to fight it to get rid of it right away! This is why some people say you can't get the same sickness twice, or if you are sick and then someone else in your family gets sick, you won't get the same sickness again from them.

There are many conditions that can occur due to cells not working the way they are meant to. Some of these that you may have heard of include cancer, diabetes, heart disease, Alzheimer's, and many others. Many common problems that come with getting older are also due to cells!

Cancer happens when cells that are not normal grow and spread very fast. Normal cells know when to stop growing and will eventually die when it's time for new cells to take their place. However, cancer cells continue to divide and divide out of control, and do not die when they are supposed to. Just like our normal cells work together in clumps to form our tissues and muscles, cancer cells work together to form tumors, or clumps of unhealthy cells. Tumors can destroy the normal cells around it, which eventually damage the body's healthy working tissues, muscles, and organs. Sometimes cancer cells will break away from the tumor and travel to other parts of the bodies and create new tumors. This is called metastasis (muh-tass-tuh-sis) and is how cancer spreads.

Cancer is not contagious. It is not spread by germs like getting a cold or the flu. It is safe to talk to, play with, and hug someone with cancer.

Doctors are not always sure why people get cancer. Some people who are perfectly healthy and take care of their bodies exactly like they are supposed to will still get cancer, and there is no reason why. Some cancers are more likely to occur from unhealthy habits. Skin cancer can come from not wearing any sunscreen over and over and always getting sunburnt. But it also depends on the person – some people's bodies are more sensitive and may be more likely to get cancer if they don't wear sunscreen. A lot of smoking or vaping can cause lung cancer. If you don't exercise and eat healthy, your cells don't have the nutrients they need to do their jobs, which can lead to different types of cancer. Sometimes cancer can be hereditary, which means if there is someone in your family that has it, there is a better chance that you could get it. This doesn't mean that you will, but it might mean that you should get checked by a doctor more just in case.

Fun Fact

Cancer is very rare in kids – only about 1% of all cancer patients in America are children.

Cancer is not yet curable, but some types of cancers can be treated. The way that doctors will treat cancer depends on what type of cancer it is and the size and number of tumors that are present. One way to treat cancer is through surgery. During surgery, doctors will try to take out the tumor and as many cancer cells as possible. They also might need to take some healthy cells that are around the cancer area to make sure they got it all.

Another type of cancer treatment is called chemotherapy, which a lot of people just call chemo (kee-mo) for short. Chemo is when the person with cancer takes a special type of medicine that attacks the cancer cells. It is usually done through an IV, which is a bag that holds the medicine connected to a tube that is connected to a needle which goes into the person's vein. This way the medicine can go straight to where it is needed. Usually if someone is doing chemotherapy, they have to do it multiple times every week for weeks or months. These drugs are very good at destroying cancer cells, but unfortunately, they also destroy healthy cells in the process. This can cause problems like tiredness, vomiting, and hair loss. This is why some people with cancer don't have hair – it is not from the cancer itself, but from the chemotherapy treatment. After treatment is over, their hair grows back.

The last common form of treatment for cancer is radiation therapy. This is like a special type of X-ray that uses high-energy waves to damage and destroy cancer cells. It is one of the most common treatments for cancer and will often be tried first before surgery or chemotherapy. It has the same side effects as chemo, which include tiredness, vomiting, and hair loss.

Sometimes even after the cancer is gone, doctors will continue treatment for a little while to make sure the cancer cells do not come back. There is no cure to cancer, which means sometimes these treatments do not work, and doctors cannot get rid of the cancer. Sometimes doctors cannot treat cancer if the tumor is in a dangerous area to treat, such as the brain or spine. Sometimes the treatments themselves can put a person at risk because they are more likely to get other illnesses while their body is weak during cancer treatments.

Diabetes is another condition that happens when your cells aren't working properly. You have an organ in your body called a pancreas. Its job is to make insulin, which helps the food you eat get to your cells. If your pancreas isn't working correctly, the nutrients in your food can get stuck in your blood, which makes your blood sugar high. This is diabetes.

The most common types of diabetes are type 1 and type 2. With type 1 diabetes, your pancreas is unable to make insulin. Most people who have type 1 diabetes find out when they are young and have to give themselves insulin every day to stay alive. They often have a little device that will tell them what their blood sugar level is at so that they can keep it at a normal amount. Most people with diabetes have type 2. Type 2 diabetes is when the cells in the pancreas start to put up a wall that doesn't let insulin in. The body must start working harder to make more insulin, and eventually can't make anymore. People with type 2 diabetes may need to take shots, and they can take medicine to help insulin get into the body's cells. Many people use insulin pumps to automatically give them insulin so they don't need shots all the time.

> ***Fun Fact***
>
> *Diabetes was discovered by Thomas Willis in 1674... by drinking urine! He sampled the urine of the patients who had similar symptoms and thought it was unusual that their urine was "wonderfully sweet like sugar or honey." This helped many scientists learn more about what diabetes was and find ways to treat it.*

Symptoms of diabetes include being thirstier than normal, losing weight for no reason, increased hunger, being very tired, body not healing quickly/normally from minor cuts and bruises, numbness in hands or feet, blurred vision, having to urinate more often, and dry skin. Some people only have a few of these symptoms, while others may have all of them. The cause of type 1 diabetes is mostly unknown. Type 2 diabetes can be hereditary (or passed down from parents), or be caused from being overweight, not exercising regularly, eating lots of unhealthy and sugary foods, and consistently not getting enough sleep. Although you can die if you don't get treatment for diabetes, most diabetic people live long and normal lives.

The more you take care of yourself, the more your cells can take care of you to help you to live a healthy life. Eating healthy, drinking plenty of water, exercising, getting enough sleep, wearing sunscreen, brushing your teeth, washing your hands and body regularly, and avoiding smoking/vaping are some ways that you can help your cells to do their job properly.

Heart attacks can be caused by unhealthy eating. When your blood cells are carrying oxygen and nutrients to the rest of your body, it is sometimes carrying fatty substances for your body to use as energy. However, these fatty substances can start to clog your veins. They might create a small layer all around these tubes that your blood travels through, so that your blood has less space to get through. If there is enough of this fat buildup, it can block the blood completely so it can't get oxygen to certain areas that need it, like parts of the heart. The parts of the heart that can't get oxygen may stop working and this could create heart attacks, high blood pressure, and other problems. Eating a variety of healthy foods and exercising regularly can help to stop this from happening!

Fun Fact

Your heart cells can replace themselves, but if you have damage within your heart cells, they cannot be replaced by the heart cells of another person.

Sometimes issues with our cells can come from accidents. A common problem that involves the cells in your brain is called a *concussion.* This is a brain injury that happens when you hit your head very hard. It is more common with athletes like football players, who often have head-on collisions with other athletes. The cells that make up your brain are protected by fluid and your skull. A small bump to the head normally won't harm your brain because of these cushions. However, if you hit your head hard enough, you can get a concussion from your brain bumping into your skull. People who have a concussion might feel dizzy, have blurred vision, get a bad headache, be confused, or throw up. These symptoms won't last forever and it's important to get lots of rest after a concussion to let your brain cells heal and repair themselves. If you have multiple concussions, especially in a short amount of time, you can get permanent damage to your brain. Some effects of this could include memory problems, "foggy thinking," trouble with focusing, and even eyesight issues.

Fun Fact

After you have had one concussion, you are more likely to get a second one!

Humans have white blood cells, but what do plants have to protect them from disease? Their cells have cell walls, which work as an extra layer of protection to keep bacteria and viruses out. Some plants have bark, which has many layers, to keep insects out. Some plants have thorns to protect themselves from being eaten by animals. Some plants release poisonous chemicals to make the animal eating it sick, so it will learn not to eat it anymore. Plants have to find unique ways to defend themselves since they are not able to just get up and walk away from predators!

Chapter 8: Single-Celled Organisms

An **organism** means a living thing. Some organisms, like plants and animals, are made up of trillions of cells. A single human cell is alive, but it is not an organism because it cannot live on its own. It needs other cells to work together to get food and create tissues, muscles, and organs. For example, if you scraped off one of your cheek cells, it would not be able to get food to grow and reproduce so it cannot stay alive. However, there are some organisms that are just a single cell. They live on their own and get their own food. Most cells are so small that they can't be seen unless you use a microscope. This is true for single-celled organisms as well.

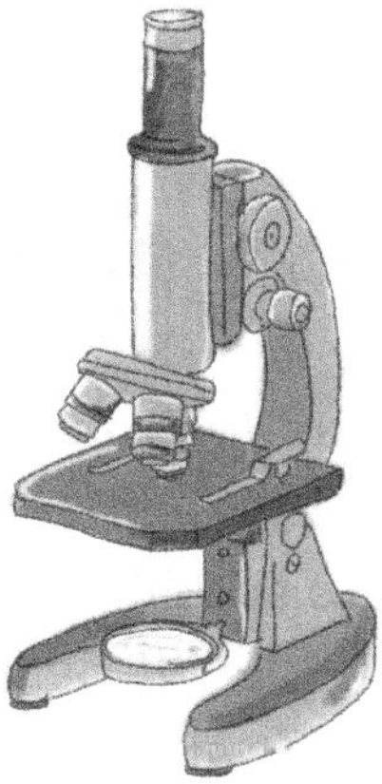

A popular type of single-celled organism is called a protist (pro-teest). They are not considered an animal, plant, bacteria or fungi. Basically, any organism that doesn't fit into one of those categories is considered a protist. Just like in a human cell, protists have organelles like vacuoles, cell membrane, and mitochondria. Protists also live near or in water.

One type of protist that you may have heard of is called algae (al – gee). You can find algae in ponds, lakes, or the ocean. Some people refer to it as "pond scum" because it can look like a frothy green layer of filth on the top of ponds. Many people think that algae is a plant, but it is actually a single-celled organism. It does have similarities to plants, like how they both use photosynthesis to get energy. They are not considered plants because they don't have leaves, roots, and stems. Algae is also the green stuff that you might see growing on the side of a dirty fish tank or swimming pool. Other types of algae that are multicellular include seaweed and kelp. Algae might seem like one big plant because it can grow together to make up the green stuff you see in a pond. It is like a snowball – it looks like one big chunk of snow, but if you look closely, you can see that it is actually made up of tons of tiny, individual snowflakes.

Amoebas (uh-mee-buhs) are another type of protist that you might be familiar with! They can only be seen under a microscope. They don't have a fixed body shape, so they kind of just look like floating blobs, constantly changing shape as they move. Amoebas don't have mouths, so they float around looking for food, like small bits of algae or tiny bits of dead plants and animals. When they come across these, they wrap their bodies around it to suck it in and absorb it. Then its organelles get to work to take the nutrients out of it that they need, just like when humans eat! They also reproduce by cell division, or basically splitting into two.

Fun Fact

If you cut an amoeba in half, the half that has the nucleus will survive! The other half will not.

Paramecium are another type of protist. It's somewhat unbelievable to think that something so small, made of one cell and no organs like animals, can be a living thing. However, it is a living thing because it has all of the seven traits that are required to be considered alive. Paramecium are covered in tiny hairs called **cilia** that help them move in the water. They do not have mouths to eat like animals. Instead, they have a small fold on one side of them called an **oral groove**. They use their cilia to help push food into this groove, which will fold over it into a vacuole for the cell to pick out the nutrients. Any waste is dumped out of the cell through the membrane. This is similar to how an animal's digestive system will pick out the nutrients in food and then dispose of what it doesn't need. Paramecium are important to ponds and lakes because they eat bits of dead organisms as food, which helps keep the water clean.

There are also single-celled organisms that are not protists, including bacteria. When you hear the word bacteria, you often think of germs or something that will get you sick. However, there are actually tons of different kinds of bacteria, and some of them are needed by your body to survive! Bacteria are single-celled organisms, which means they are a living thing. Unlike most cells, bacteria do not have a nucleus to tell them what to do. They come in all shapes and sizes, including rods, spirals, and spheres. Most bacteria move around using something called a flagellum, which is like a tail. They reproduce similar to protists, in that they split into two to basically make a clone of themselves.

Bacteria are found everywhere. They are in and on every living thing. Bacteria are one of the few living things that can survive in almost any condition – boiling hot, freezing cold, wet, dry. They are in the air, in water and ice, in dirt, in food, in dust and rocks, and even the ocean. There is more bacteria than any other organism on Earth. It's impossible to know how many bacteria or even how many different kinds of bacteria there are.

Fun Fact

Scientists estimate that there are around 100 times more bacteria in the ocean than there are stars in the universe!

Bacteria play an important role in the Earth. You have bacteria in your stomach that helps break down the food that you eat. These are helpful. Other helpful bacteria might live in the soil and help decompose dead plants/animals. This helps the soil to be fertile and gives it food. Other bacteria are not good or helpful. These are called pathogens. Rotten meat and other foods, especially if they haven't been cooked all the way, may have bacteria that can make you sick with food poisoning. Thoroughly cooking and washing your food can help remove bacteria that is harmful to humans. There is also a type of bacteria that can give you infections like strep throat or Lyme disease.

Fun Fact

Cheese, yogurt, and pickles are created using special types of bacteria!

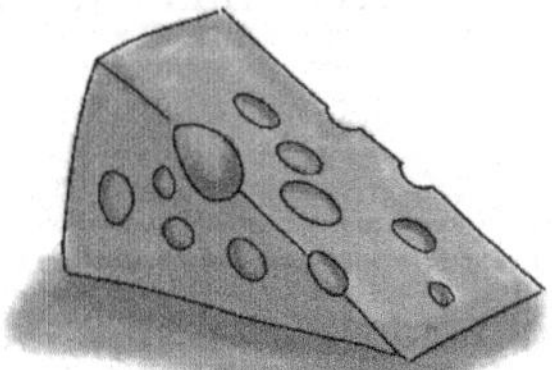

A virus is not a living thing. It is not made of cells. Instead, it has DNA or RNA (which is similar to DNA) protected by a layer of protein called a capsid. It cannot reproduce on its own, which is the main reason it is not considered a living thing. Instead, it finds a cell and will inject itself inside of the cell and infect it. They are not active unless they are inside another living cell. They can float through the air or survive in water or on the skin. You can get a virus from another person by shaking hands with someone who has it, through water like a swimming pool or lake, and through the air after an infected person coughs or sneezes. That's why it's so important to always cover your mouth with your elbow when you cough or sneeze – to prevent the spread of viruses! You can also get viruses through infected insects, animals, or even food that has gone bad.

Fun Fact
Viruses are so small that most microscopes aren't even powerful enough to see them!

Viruses cause many illnesses, such as the flu, smallpox, rabies, warts, and the chicken pox. It is hard to find medication to fight viruses, because most medicine only targets bacteria. The best way to fight a virus is to get a vaccination before the virus enters your body. Most babies and kids will get a vaccination, or shot, for common sicknesses like the chicken pox. These shots contain tiny bits of the virus so that your body knows what it looks like and can figure out how to fight it, without actually making you sick. This way if you do get exposed to the virus later in life, your body will already know how to fight and kill it! Polio and smallpox disease used to kill a lot of children, and now because of vaccinations, these diseases have almost completely disappeared! Some people will get a flu shot every year. The reason you have to get more than one is because there are different types of flu each year. The flu from last year might not be the same as the flu that is going around this year, so the vaccination wouldn't be effective anymore. Similar to bacteria, not all viruses are bad either. For example, some viruses can kill bacteria, including bacteria that are harmful to humans.

Chapter 9: History

The history of how cells were discovered starts with scientists questioning where life comes from. Before the microscope was invented, no one could imagine something being alive that they couldn't see. After the microscope was invented, scientists discovered cells and eventually learned what they were made of and how they worked. Now they use this information to figure out how to prevent/cure illnesses, improve crops, and other ways to improve the world!

It all started in the 300s BC when the Greek philosopher Aristotle wrote about the theory of spontaneous generation (life created from nothing). Everyone accepted this as true information, and it took over 2,000 years for scientists to discover that this theory is incorrect!

In 1590, Hans and Zacharias Janssen accidentally created the first

> ***Fun Fact***
>
> *Prison cells also get their name from the Latin name Cellula and may look similar to what Rober Hooke first saw in the cork. Prison cells are really just small rooms or compartments stacked by each other.*

microscope by putting two glass lenses in a tube. They were not even scientists, but eyeglasses makers! They were able to magnify objects three to nine times bigger using this device. Cells were not discovered until around sixty years later, but this was an important first step to get there. We would not know about cells if we did not have the ability to see them using microscopes. Much later, Zacharias also invented the first compound microscope, which used two lenses.

In the 1600s, an Italian scientist named Galileo Galilei learned how to create a powerful telescope to study stars using lens grinding. He made many discoveries, including that the moon's surface is rough and uneven. Before this, everyone thought it was smooth! During his studies around 1624, he discovered that a telescope could be used to magnify insects. With this information, he was able to create a microscope that had the ability to focus. The general design of his microscope was not used by scientists (it was basically a little telescope, and not very powerful), but the way he was able to make it focus did become standard in future microscopes.

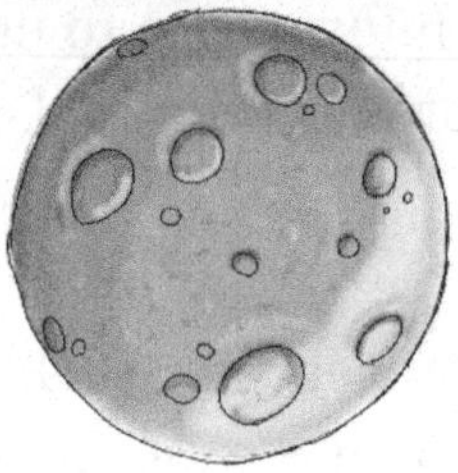

In 1660, an English physicist named Robert Hooke used the invention of the microscope to look at cork from a tree. He saw what looked like rows of empty rooms or compartments, and called them cells. This comes from the Latin name Cellula, which means small compartment. He is also the one that added a light to the microscope. As a child, Robert was often very sick so he spent a lot of his time schooling at home. This is how he became so interested in the world around him and what it is made of. Before becoming a scientist, Robert Hooke was an artist. This became a very useful skill for him as a scientist, because he was able to use his artistic talent to make scientific drawings of his discoveries. He wrote a book called *Micrographia* in September 1665 about all of his discoveries.

Around the same time, a Dutch microscopist named Antony van Leevenhook became the first person to see living cells and discovered single cellular organisms. He discovered this by observing the bacteria from plaque that he scraped off his teeth! Antony experimented and invented ways to change the lenses on a microscope so that they would increase the magnification. For example, he ground and polished a small glass ball and made it into a lens to create a magnification of 270 times. With his inventions, he also discovered protists when he reported seeing tiny objects swimming around in drops of pond water. He called them animalcules, meaning tiny animals.

Fun Fact

Antony Van Leevenhook created around 400 different microscopes during his lifetime!

A couple years later in 1675, the Italian scientist Marcello Malpighi published several papers and books on the early development of living things. He mostly studied the way that a chick and insects grow inside of an egg. He is considered to be the "father of embryology". Embryology is the study of cells that are growing to create a baby.

About 150 years later in 1831, the Scottish biologist Robert Brown found that the nucleus is a key part of each cell. This is where DNA is stored. Other scientists had seen it before, but never thought it to be very important. He mostly studied plant cells, and once he sailed from England to all over Australia to collect and study different types of plants. His collection included around 4,000 species that were almost all new to science!

A few years later in 1839, German scientists Theodor Schwann and Matthias Schleiden stated that all life on Earth is composed of cells. Schwann studied tissues from lots of different animals under his microscope and discovered that they were all made of cells. Schleiden looked at lots of different plants and parts of plants under his microscope and discovered that they were all made up of cells. They compared the differences in their findings between animals and plant cells and decided that all living things must be made of cells. They called this discovery the "Cell Theory of Life", and it is still used today.

About twenty years later in the 1860s, a German scientist Walther Flemming used dye to stain cells to see them easier under a microscope. This method led to his discovery of how cells reproduce by splitting. At the same time, the Austrian monk Gregor Mendel used pea plants to show how genes work. He experimented by combining different types of pea plants to create new ones that were a mixture of their "parents". At first he was ignored for years, but by 19000 his work became the foundation of genetics, also known as the study of genes.

Forty years later in 1900, German physician Rudolf Virchow stated that all cells come from existing live cells. He discovered this by studying lots of different diseases and what caused them. This is the first time that scientists realized Aristotle's theory of spontaneous generation (life created from nothing) was incorrect.

In 1931, German physicists named Ernst Ruska and Max Knoll invented the electron microscope, which made it possible for scientists to magnify objects up to a million times. This made it possible to see even atoms! This gave scientists much greater insight on the inner works of cells and their organelles. Ruska won a Nobel Prize in Physics for his work.

About twenty years later in 1953, American biologists James Watson, Francis Crick, and Maurice Wilkins discovered DNA, which led to many breakthroughs in genetics. They created the model of DNA that looks like a double helix (which looks like a twisting ladder), which is still used today. They were awarded with a Nobel Prize for this discovery in 1962.

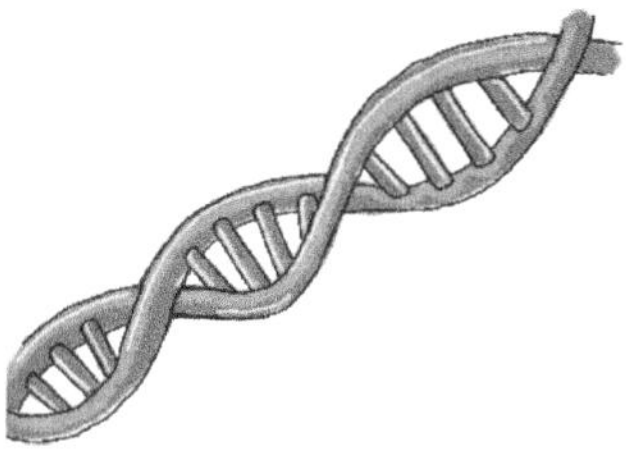

Fun Fact

James Watson also taught biology at Harvard University!

The ribosome was discovered in 1974 by Belgian chemists Albert Claude, Christian de Duve, and George Emil Palade. They found a way to break open a cell to be able to study the organelles inside more easily. They won a Nobel Prize for their discovery.

In 1981, German physicists Gerd Binning and Heinrich Rohrer invented the scanning tunneling microscope. This allowed scientists to look at objects down to the atomic level in three-dimensional images. They won the Nobel Prize in Physics in 1986, and it is still one of the most powerful microscopes today.

In 2003, the Human Genome Project finished a map of the entire human genome (all genes found in the DNA of a cell's nucleus). Before, scientists only knew about 90% of the human genome. This is one of the greatest scientific achievements in history. Scientists from the United States, United Kingdom, France, Germany, Japan, and China all came together to work on this project. It took about thirteen years to complete. When they finished this project, they immediately shared every part of it to all scientists and shared their findings at many conferences. The findings from this project helps doctors and scientists to predict, diagnose, and treat illnesses at the cellular level, where they begin. For example, Huntington's Disease is a genetic condition that is commonly passed from a parent to their child. Its discovery was during genome mapping before the Human Genome Project, and was one of the big motivations for the project to get started. Now that it has been found in DNA, scientists can do genetic testing to make sure that the disease isn't passed on to future children.

As you can see, discoveries don't happen overnight with just one scientist. Scientists all over the world slowly make discoveries over time through a lot of experimentation. Theories and "facts" are constantly changing as scientists continue to explore. Many scientists will get a job as a professor to share their knowledge and work with their students to keep doing research and experimentations. There are also many conferences where scientists who study the same thing, such as cells, come together from all over the world to share their findings with each other. They can also compare their notes to see if they have similar findings and theories. They do this so that they can learn from what others have found, which may help their own work and lead to new discoveries.

Made in the USA
Monee, IL
04 November 2026

67383964R00036